VIRGO

Published 2025

FiNGERPRINT!
Prakash Books

[f] Fingerprint Publishing
[X] @FingerprintP
[O] @fingerprintpublishingbooks
www.fingerprintpublishing.com

ISBN: 978 93 6214 656 4

Contents

Who Are We?

Born between 23 August
and 22 September, we,
the Virgos, are the ultimate mix
of meticulousness and modesty.

We're known for our
analytical minds, attention
to detail and a strong
desire to help others.

We're the ones who'll
notice the smudge on your
glasses or find that one missing
sock you've been searching for.

Virgos are ruled by Mercury,
the planet of communication,
which might explain our knack
for organizing our thoughts
and expressing ideas clearly.

We tend to be practical,
grounded and dependable,
though sometimes we can
be a bit too self-critical due
to our perfectionism.

Virgo's Virtuous Voyage

The Beginning of Our Story

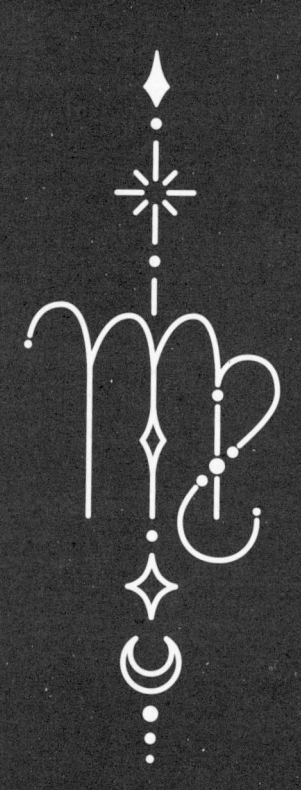

Virgos' journey, from ancient mythology to the modern times, reveals a rich tapestry of symbolism and traits, showcasing our timeless appeal and significance.

From the nurturing goddesses of antiquity to the meticulous individuals of today, we have always been associated with qualities of care, precision and intellectual depth.

First Stop: Greek Myth!

The most significant figure associated with the Virgo sign is Persephone who was the daughter of the goddess of harvest and agriculture Demeter.

Persephone was abducted by Hades, the king of the underworld.

Her absence and eventual return are symbolized by the changing of seasons, reflecting the Virgos' trait of being compassionate caregivers.

Here Comes the Roman Myth!

The Roman equivalent of Demeter, Ceres, is closely linked to the Virgo constellation.

The grain goddess Ceres embodies the sign's nurturing aspects.

Interestingly, the word "cereal" is derived from her name!

Christian Myth Ahead!

In Christian symbolism,
the sign is often associated
with the Virgin Mary, representing
purity, devotion and the nurturing
mother archetype.

*This connection emphasizes our
qualities of innocence and care.*

Zodiacs without Astrology?
No Way!

The Virgo constellation
is one of the largest in the
sky and is depicted as a maiden
holding a sheaf of wheat.

*This imagery aligns
with Virgo's association with
harvest and fertility.*

We're known for our
analytical and detail-oriented
nature, which is believed
to be influenced by the sign's
ruling planet, Mercury.

*In astrology,
we are seen as practical,
methodical and perfectionistic.*

During the medieval period, Virgos were associated with purity and chastity.

Astrologers associated us with the idea of wisdom and healing, reflecting our analytical and nurturing qualities.

Virgo Unplugged

Our Eccentric Side

You might find us color-coding our grocery list or creating a spreadsheet for our vacation itinerary.

We view life as a well-oiled machine and make it more efficient by ensuring that everything is organized.

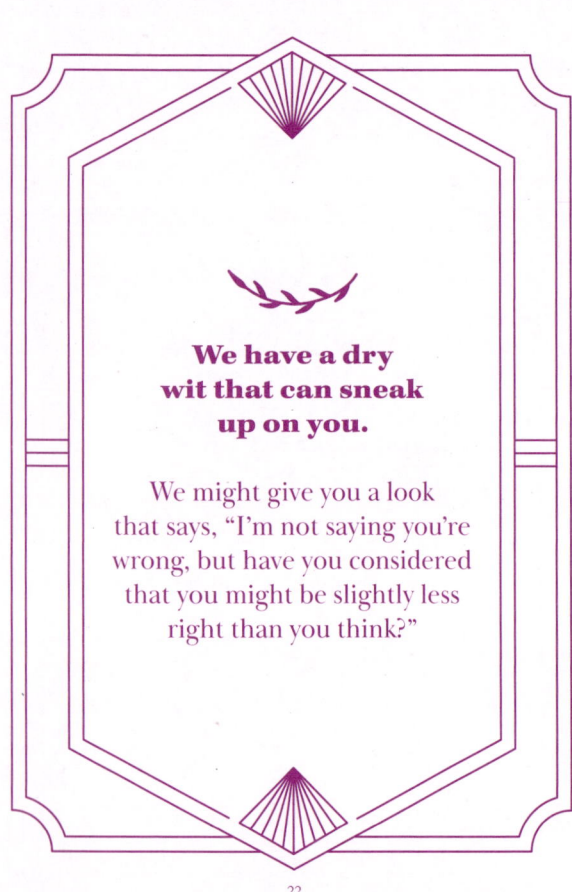

**We have a dry
wit that can sneak
up on you.**

We might give you a look
that says, "I'm not saying you're
wrong, but have you considered
that you might be slightly less
right than you think?"

**Underneath our
meticulously crafted
hardbound exterior
is a heart of gold.**

We're often generous
and quietly supportive,
but prefer to show it through
practical acts rather
than grand gestures.

**Despite our grounded
nature, we have rich
inner worlds with
imaginative dreams.**

We might surprise you
with our deep reflections
on the meaning of life—often
during another thoroughly
planned dinner party.

Behind our calm demeanor, we sometimes wrestle with intrinsically placed high standards and self-criticism.

We often are our own harshest critics, but that's because we genuinely care about getting things right.

We are detectives
in disguise.

We'll find that one
missing sock or solve the mystery
of who ate the last cookie—without
you even knowing that we had
taken on the case.

We may not shout
that we care, but we're
the ones who show up
with homemade soup
when you're sick.

Beneath our practical exterior, we have a poetic side, often reflected in our detailed and thoughtful writing or personal reflections.

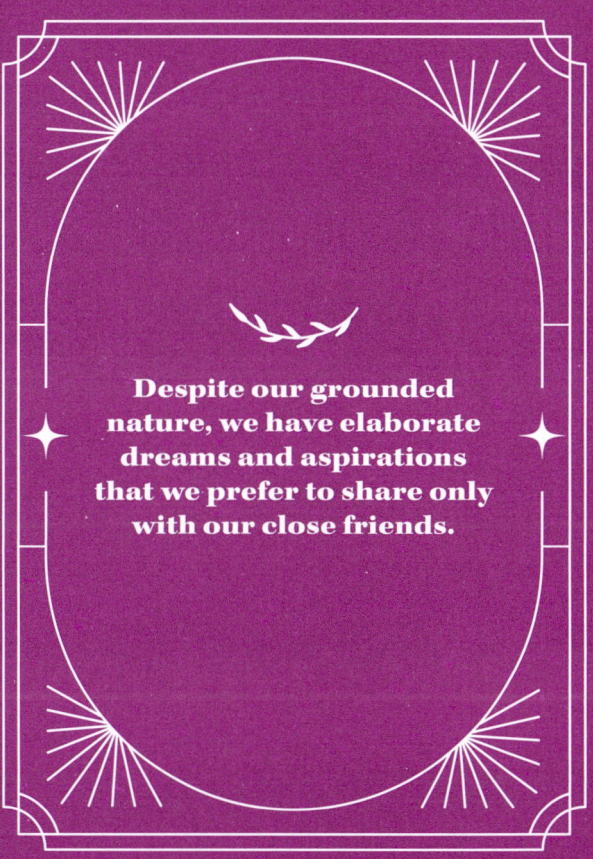

Despite our grounded nature, we have elaborate dreams and aspirations that we prefer to share only with our close friends.

Wondered What VIRGO Stands For?

♍ **V** for Virtuous

♍ **I** for Intelligent

♍ **R** for Radiant

♍ **G** for Generous

♍ **O** for Optimistic

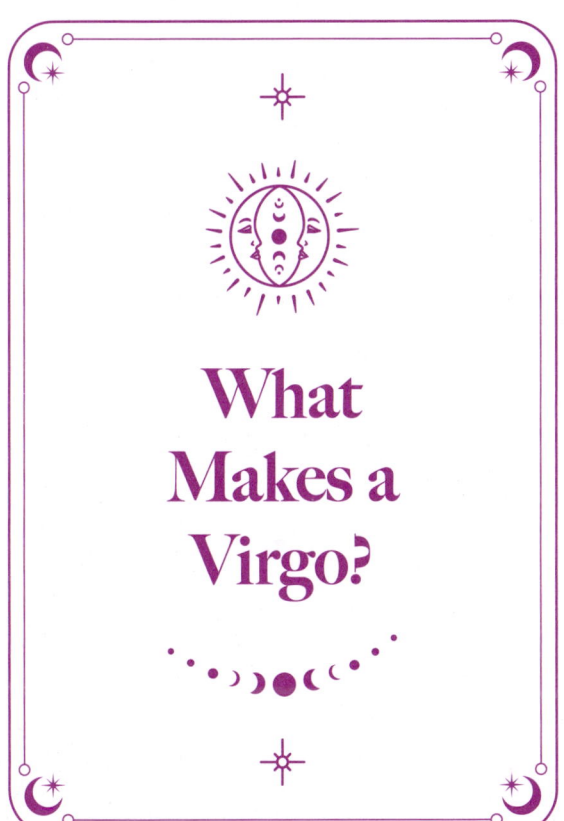

What Makes a Virgo?

We are meticulous,
reliable and have practical
problem-solving skills.
We also offer steadfast
support and care to others and
combine intellectual curiosity
with a compassionate,
grounded approach to life.

*Let's explore the qualities Virgos
are known for!*

We're known for our
exceptional attention
to detail.

We notice things that
others might miss and bring
a high level of precision
to our tasks.

We often take on a
caring role, providing
support and advice to
friends and family with
a nurturing touch.

We're masters
of organization—whether
it's managing a household,
setting up a workspace
or planning an event,
we excel in creating
order and structure.

We possess a sharp
intellect and a keen sense
of insight, making us adept
at understanding complex
situations and offering
thoughtful perspectives.

We approach life with
a practical mindset,
preferring realistic
solutions over lofty ideals,
focusing on what works
best in the real world.

We might be reserved in expressing our emotions, but we're deeply compassionate and open up to those we trust.

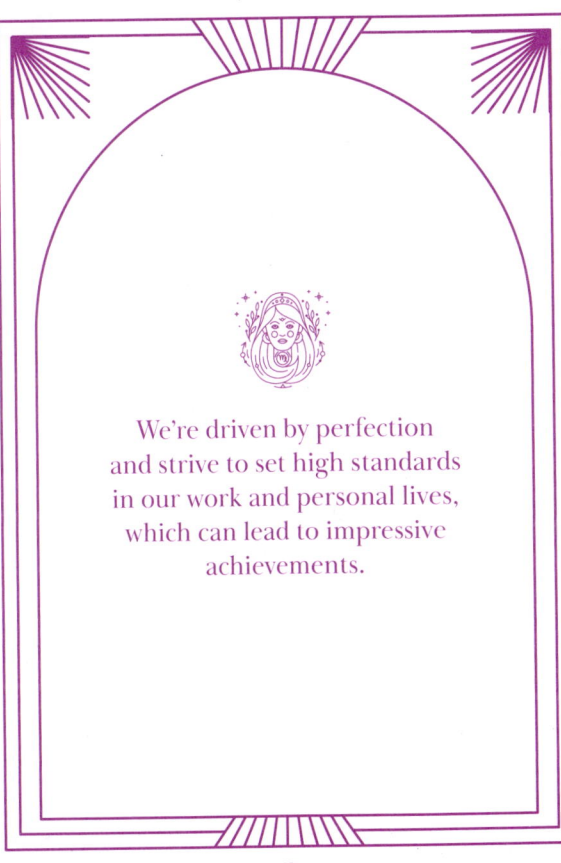

We're driven by perfection
and strive to set high standards
in our work and personal lives,
which can lead to impressive
achievements.

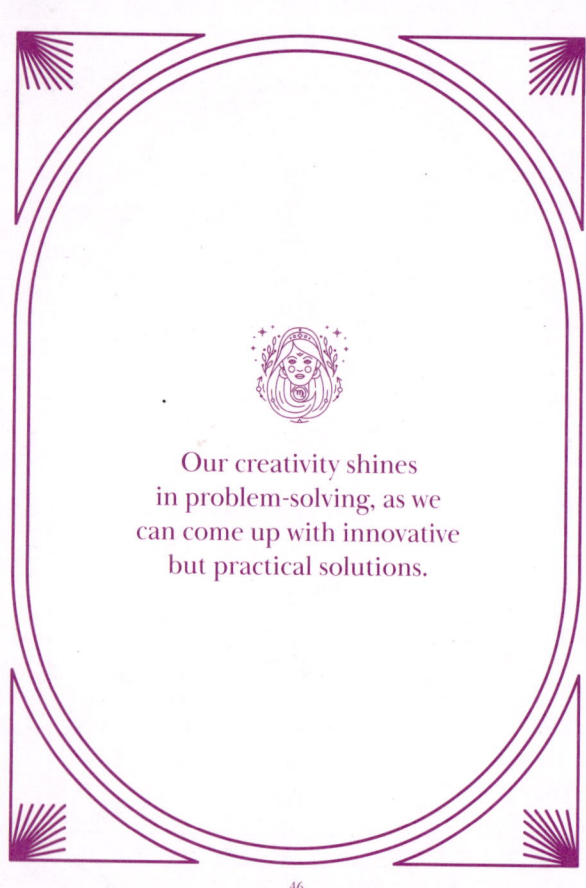

Our creativity shines
in problem-solving, as we
can come up with innovative
but practical solutions.

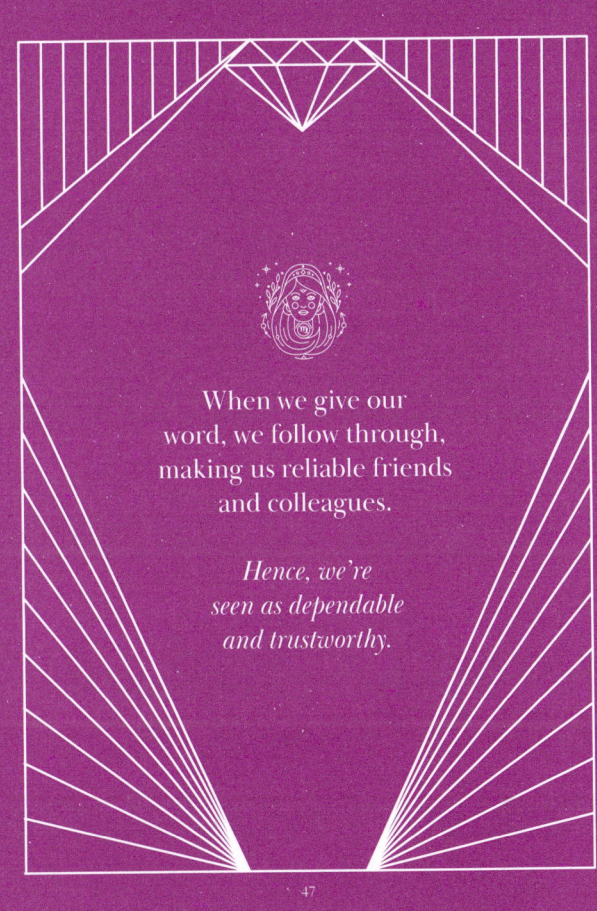

When we give our
word, we follow through,
making us reliable friends
and colleagues.

*Hence, we're
seen as dependable
and trustworthy.*

We often focus
on personal growth
and self-improvement, always
seeking to better ourselves and
refine our skills and abilities.

We often pursue knowledge
and enjoy engaging in
intellectual discussions.

*We have a natural curiosity
and love for learning.*

Our attention to detail
extends to our preferences,
resulting in refined tastes and
a good eye for quality in things
like fashion or hobbies.

We're excellent
at troubleshooting and
finding logical solutions, as we're
often the go-to person when
a problem needs fixing.

We're our toughest critics,
striving for perfection
and setting high
standards for ourselves
in everything we do.

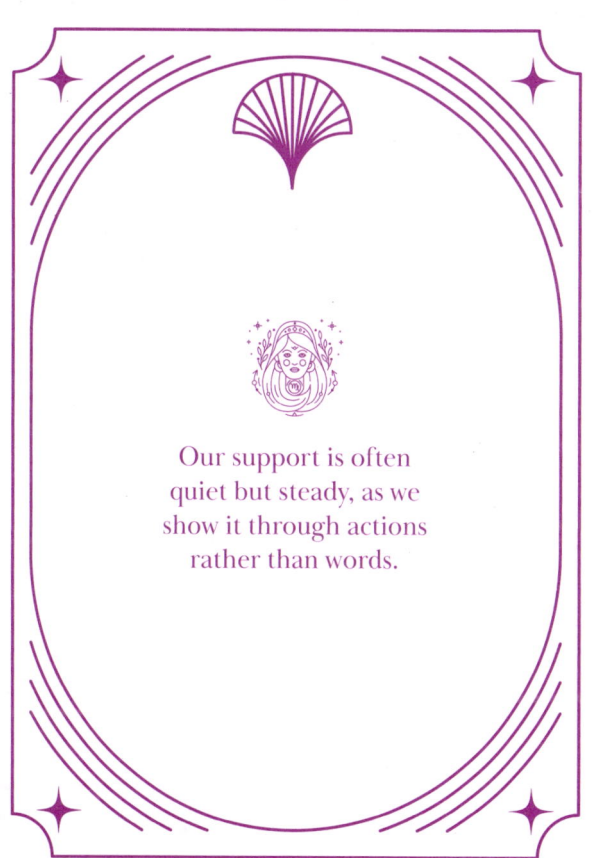

Our support is often
quiet but steady, as we
show it through actions
rather than words.

Virgo

From
Ancient Myths
to
Modern Hits

The Virgo Journey

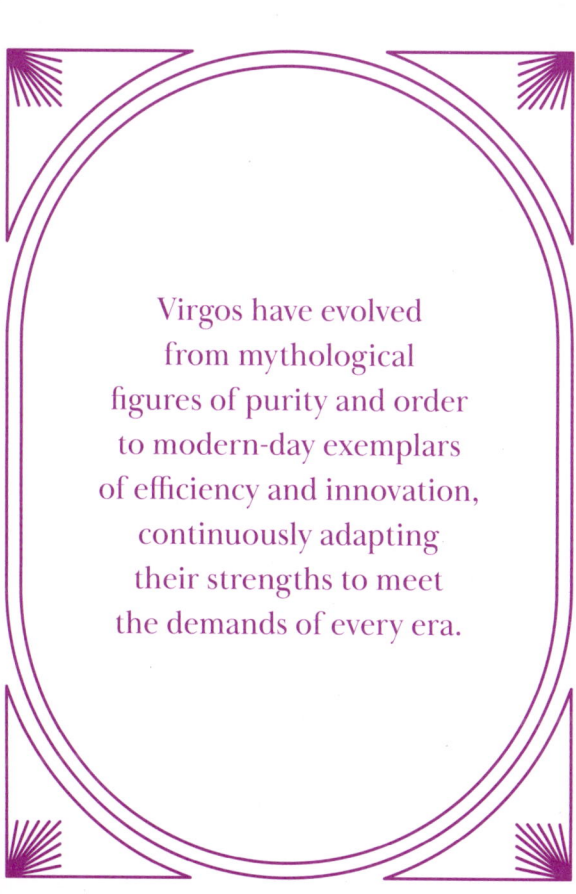

Virgos have evolved
from mythological
figures of purity and order
to modern-day exemplars
of efficiency and innovation,
continuously adapting
their strengths to meet
the demands of every era.

In ancient times,
Virgos were celebrated
as embodiments of purity
and wisdom, with their attributes
linked to agricultural deities and
nurturing figures.

Ancient cultures honored
Virgos through seasonal festivals
and rituals that celebrated harvests
and the changing of seasons,
reflecting their connection
to the fertility of the earth.

During the Renaissance,
we were admired for our
intellectual prowess and meticulous
approach to arts and sciences.

We were often seen
as scholars and reformers
who contributed to a broader
understanding of knowledge.

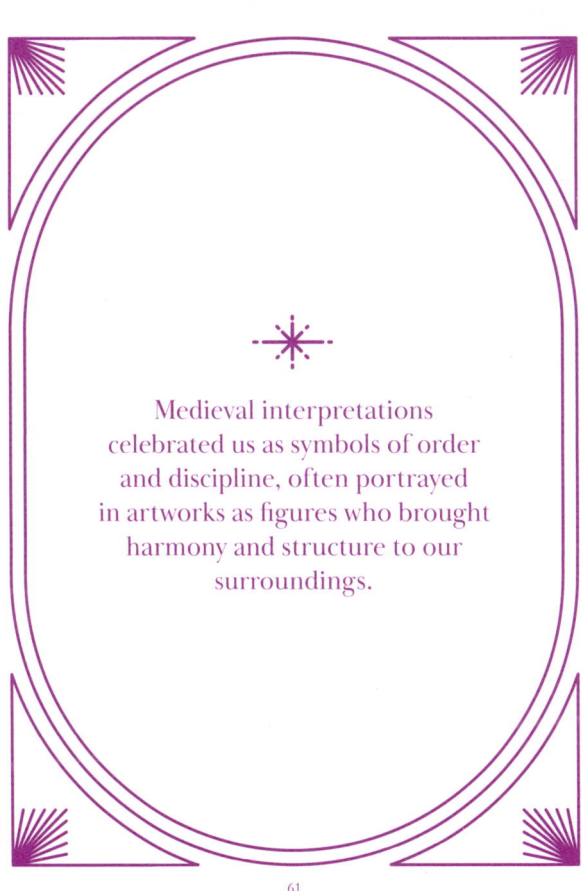

Medieval interpretations celebrated us as symbols of order and discipline, often portrayed in artworks as figures who brought harmony and structure to our surroundings.

In contemporary times, we're
recognized for our efficiency
and expertise in various fields,
from technology and business
to healthcare and education.

*Our meticulous nature makes
us sought-after professionals.*

With the advent
of technology, we've adapted
our organizational skills to
digital platforms, mastering data
management, coding and digital
communication.

*Our ability to navigate
and optimize technology
showcases our evolving role
in the modern world.*

As innovators and
problem-solvers, we leverage
our analytical skills to address
complex future challenges,
including advancements
in artificial intelligence
and environmental
sustainability.

As the focus on mental and physical well-being grows, we lead in holistic health fields, using our attention to detail and practical approach to promote a balanced and mindful living.

In recent years,
we've become cultural
influencers in the realms
of lifestyle and productivity.

Our obsession with organization
and efficiency has inspired trends
in minimalism, bullet journaling
and productivity hacks, influencing
how people approach their daily
lives and work routines.

We're increasingly
recognized as champions
of environmental sustainability.

*Our practical approach
makes us natural advocates
for eco-friendly practices
and sustainable living.*

We're often at the forefront
of green technology,
waste reduction initiatives
and environmental
conservation efforts.

Virgo in Love

The Reliable
Romantics of the Zodiac!

We approach love
with the same precision
that we apply to other
areas of our life.

We'll remember our
partners' favorite flower,
preferred restaurant and
even the little things they
mention in passing.

While we may seem
reserved, we're deeply
passionate beneath our
practical exterior.

We express love
through thoughtful gestures
and consistent support rather
than grand declarations.

We have high standards
and can be quite discerning
when choosing a partner.

We seek a relationship
that meets our ideals
of compatibility and
shared values.

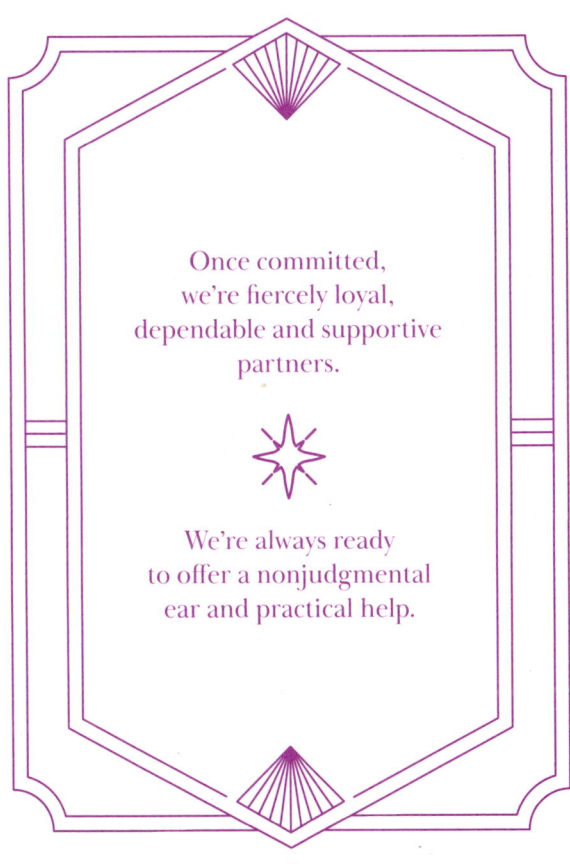

Once committed,
we're fiercely loyal,
dependable and supportive
partners.

We're always ready
to offer a nonjudgmental
ear and practical help.

*Taurus' sensuality
balances our attention
to detail.*

Ours is a grounded
and nurturing relationship,
as together we appreciate
stability and practicality.

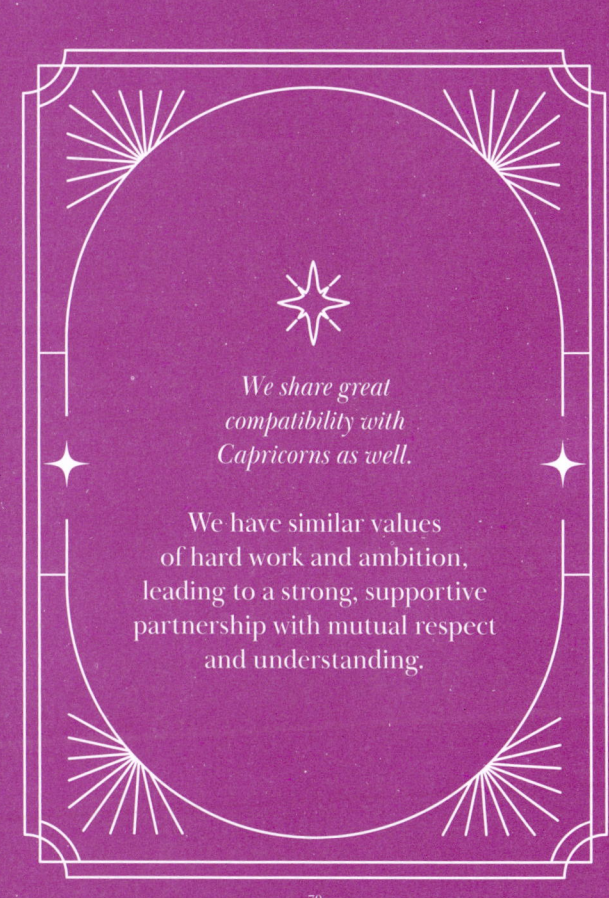

*We share great
compatibility with
Capricorns as well.*

We have similar values
of hard work and ambition,
leading to a strong, supportive
partnership with mutual respect
and understanding.

Cancer's emotional depth
complements our practical nature,
providing a balanced dynamic
where emotional needs meet
logical solutions harmoniously.

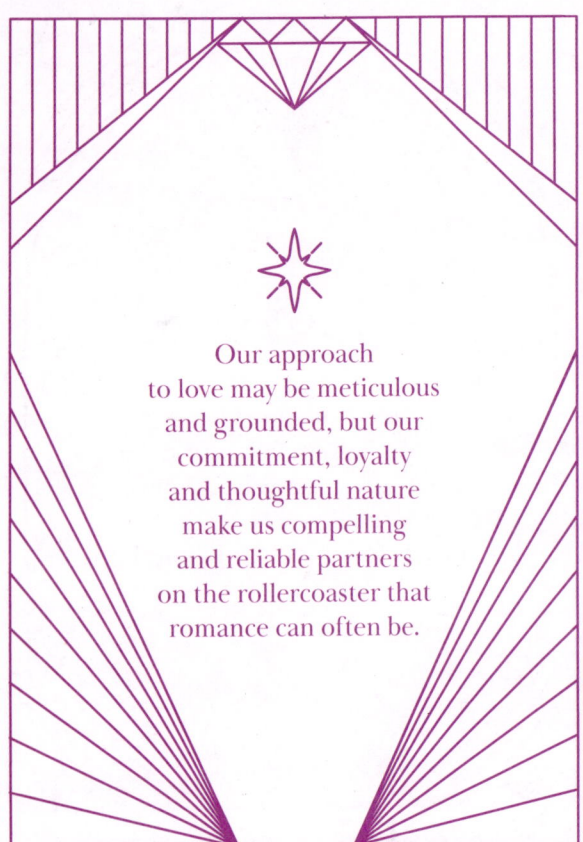

Our approach
to love may be meticulous
and grounded, but our
commitment, loyalty
and thoughtful nature
make us compelling
and reliable partners
on the rollercoaster that
romance can often be.

*We may come across
as overcritical due to our
perfectionistic tendencies.*

However, our feedback
is usually constructive
and aimed at improving
the relationship.

We prefer stability and
order in our relationships
and are not fans of drama
or unpredictability.

We value calm,
rational discussions over
emotional upheavals.

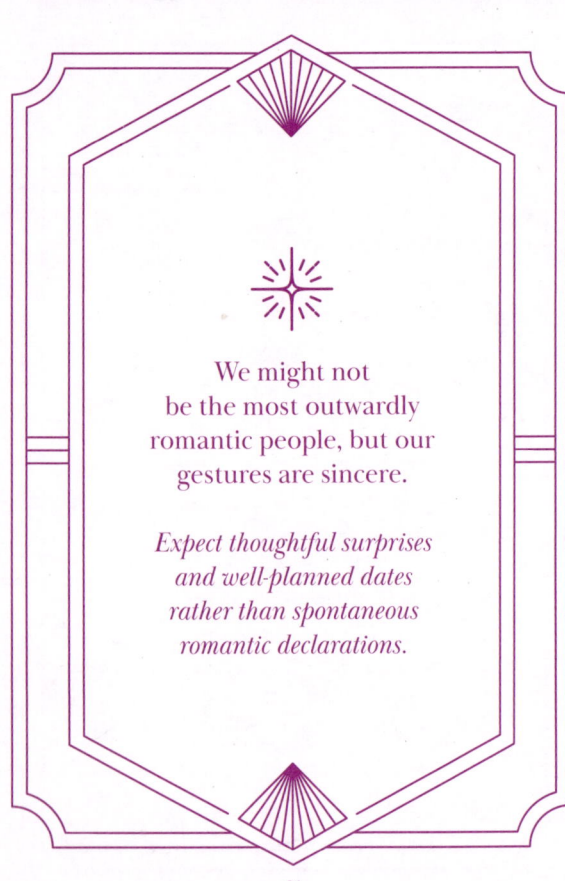

We might not
be the most outwardly
romantic people, but our
gestures are sincere.

*Expect thoughtful surprises
and well-planned dates
rather than spontaneous
romantic declarations.*

We bring a
health-conscious approach to
our relationships, promoting
wellness as part of the life
we share with our partners.

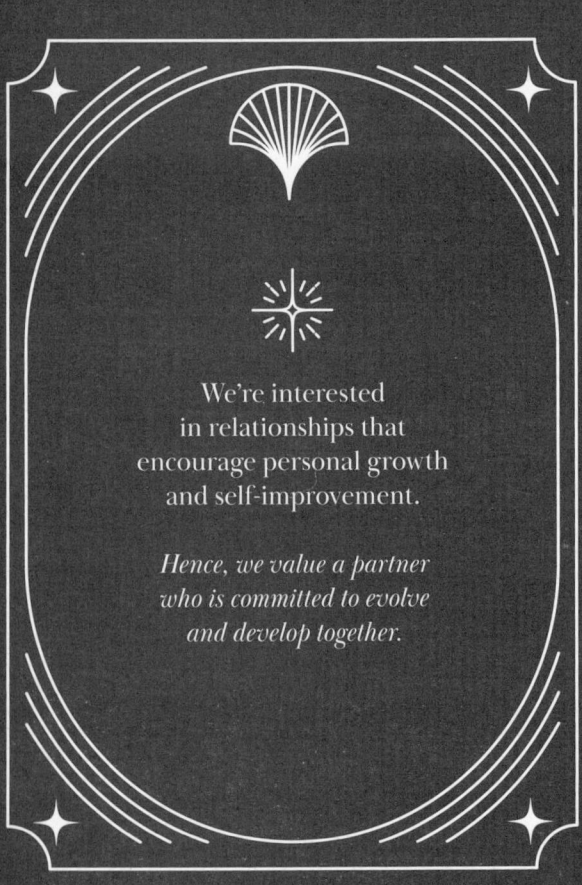

We're interested
in relationships that
encourage personal growth
and self-improvement.

*Hence, we value a partner
who is committed to evolve
and develop together.*

Stars Aligned

Who's in the
Virgo Club?

*"Time is clay.
Go make something."*

BARBARA BACH

Fashion model
and actress best known
for her role as the Bond girl
Anya Amasova in *The Spy
Who Loved Me.*

"What does fear taste like? Success. I have accomplished nothing without a little taste of fear in my mouth."

BEYONCÉ

Rose to fame in the late 1990s as the lead singer of the R&B girl group Destiny's Child.

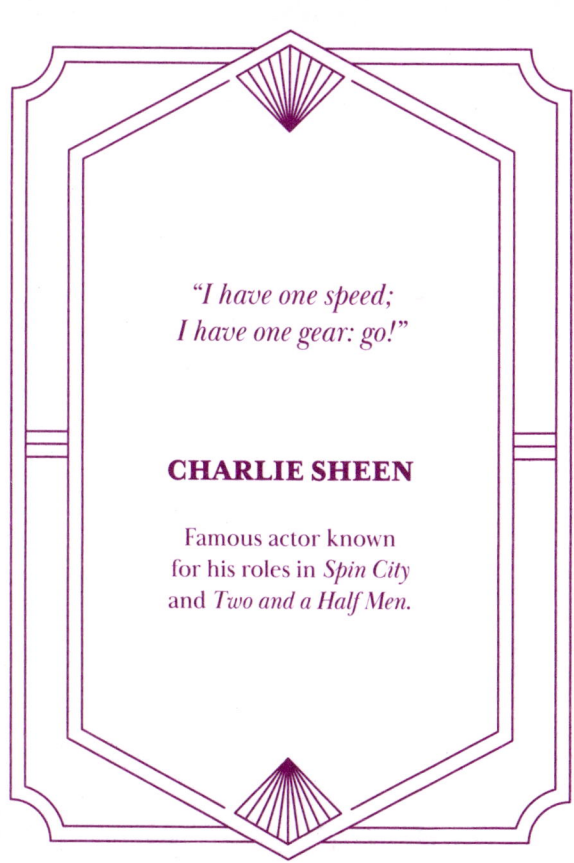

"I have one speed;
I have one gear: go!"

CHARLIE SHEEN

Famous actor known
for his roles in *Spin City*
and *Two and a Half Men.*

"You know, all that really matters is that the people you love are happy and healthy. Everything else is just sprinkles on the sundae."

PAUL WALKER

Popular actor, most known for his role as Brian O'Conner in *The Fast and the Furious* franchise.

*"I'd rather be by myself
than be spending any time
or energy on somebody that
I didn't feel sure about."*

BLAKE LIVELY

Popular actress, known
for her roles in *Gossip Girl*, *The
Sisterhood of the Travelling Pants*
and *A Simple Favor*.

*"A happy heart comes first,
then the happy face."*

SHANIA TWAIN

Known as the
"Queen of Country," and
five-time Grammy winner, she is
one of the most commercially
successful artists of all time.

*"You never know what
life has in store for you, but I
believe there are certain things
one is meant to go through."*

GLORIA ESTEFAN

Cuban-American singer,
rose to fame as the lead singer
of Miami Sound Machine, became
one of the biggest Latino musical
acts in the world by the 1980s.

*"There's nothing more
addictive or incredible in life
than reinventing yourself and allow
yourself to be different every day."*

THALÍA

One of the most
influential Mexican artists
in the world, known as the
"Queen of Latin Pop."

*"Beauty is how you feel inside,
and it reflects in your eyes."*

SOPHIA LOREN

Renowned Italian
actress, who won an
Academy Award for her film *Two
Women*, which was the first for a non-
English-language performance.

"You can be a kid as long as you want when you play baseball."

CAL RIPKEN JR.

Nicknamed "The Iron Man," played 21 seasons in the MLB for the Baltimore Orioles, and is best known for breaking Lou Gehrig's record for consecutive games played.

*"I don't regret anything.
Everything happens for
a reason—it's part of the healing
process. Life is a healing process."*

RICHARD GERE

An iconic actor,
best known for his roles in
American Gigolo, *Pretty Woman*
and *An Officer and a Gentleman*.

"I try to find a reason to laugh each day. Somehow, if you can incorporate laughter into your day, every day, it really helps. It's the little things in life that make me happy."

FAITH HILL

American country singer known for her successful albums *Faith* and *Breathe* and hit singles "This Kiss" and "The Way You Love Me."

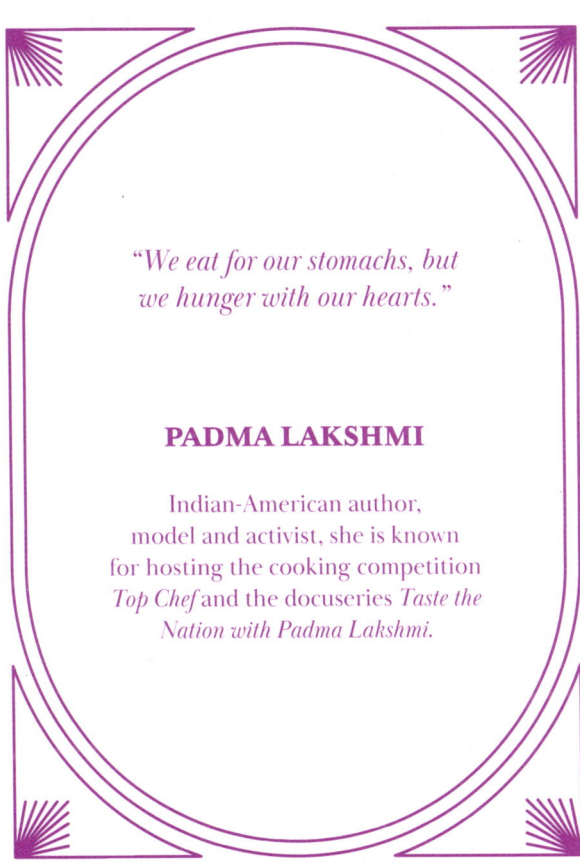

*"We eat for our stomachs, but
we hunger with our hearts."*

PADMA LAKSHMI

Indian-American author,
model and activist, she is known
for hosting the cooking competition
Top Chef and the docuseries *Taste the
Nation with Padma Lakshmi.*

*"I will always have my
songs, and I don't think
I will ever dry up."*

BARRY GIBB

Known as the falsetto
voice of the Bee Gees,
he co-founded and co-wrote
many of the group's enduring
hits along with Maurice Gibb
and Robin Gibb.

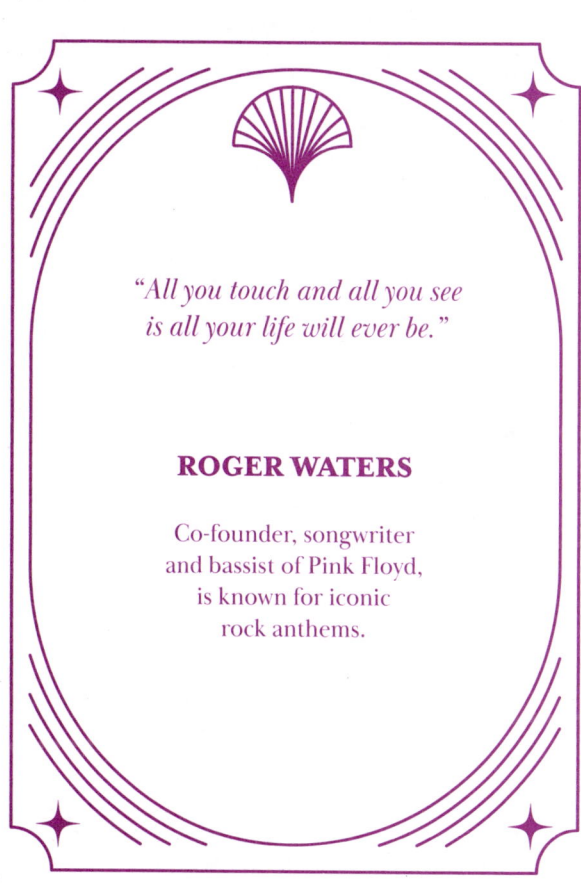

"All you touch and all you see
is all your life will ever be."

ROGER WATERS

Co-founder, songwriter
and bassist of Pink Floyd,
is known for iconic
rock anthems.

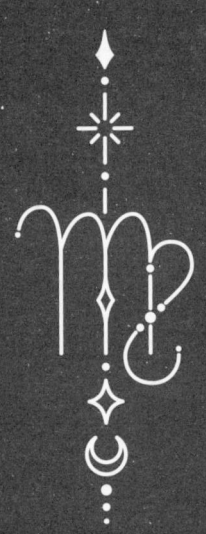

*"The point is to be
true to yourself otherwise
you may as well give up."*

KEITH FLINT

Frontman of The Prodigy,
became a significant figure
of the UK rave scene
during the 1990s.

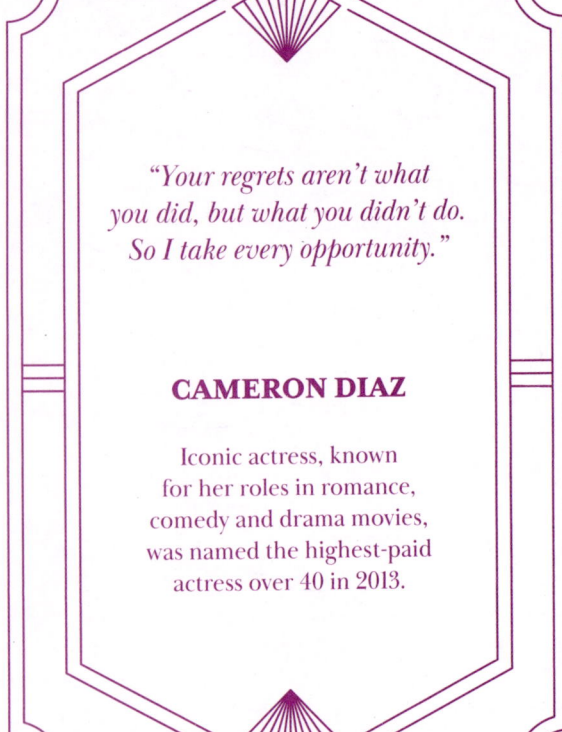

"Your regrets aren't what you did, but what you didn't do. So I take every opportunity."

CAMERON DIAZ

Iconic actress, known
for her roles in romance,
comedy and drama movies,
was named the highest-paid
actress over 40 in 2013.

*"All that counts in
life is intention."*

ANDREA BOCELLI

Italian tenor and
multi-instrumentalist who
rose to fame after winning the
newcomers' section of the 44th
Sanremo Music Festival and has since
released fifteen solo studio albums.

*"The simple act
of paying attention can
take you a long way."*

KEANU REEVES

Prolific actor with a
diverse filmography, most
known for playing Neo in
The Matrix series.

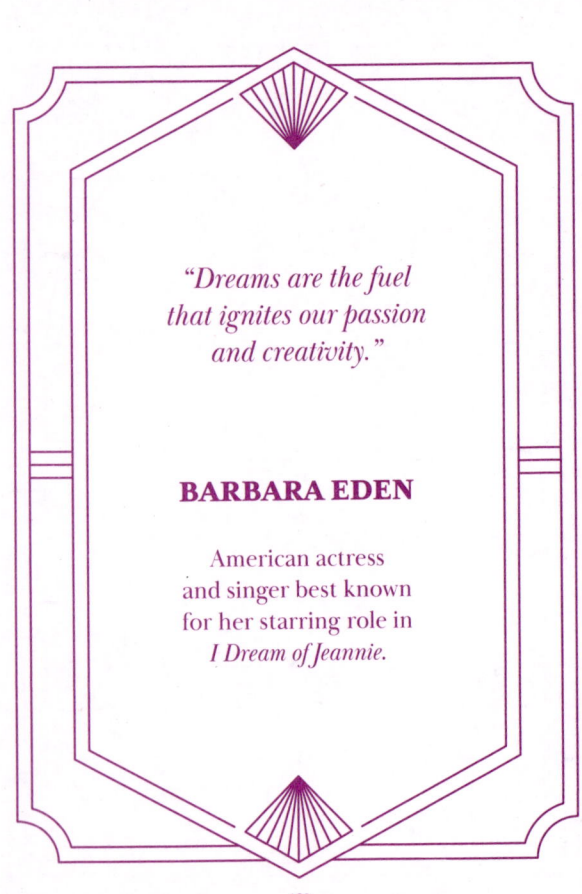

*"Dreams are the fuel
that ignites our passion
and creativity."*

BARBARA EDEN

American actress
and singer best known
for her starring role in
I Dream of Jeannie.

"I think people should have fun. And don't get so down on yourself. Enjoy life and be the best person you can be."

KEKE PALMER

American actress, singer and author, she started her career with *Barbershop 2: Back in Business.*

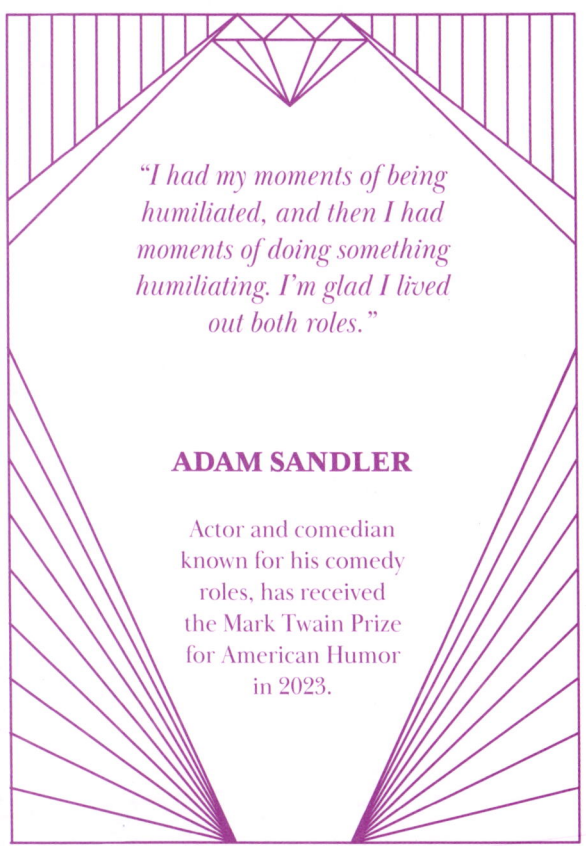

"I had my moments of being humiliated, and then I had moments of doing something humiliating. I'm glad I lived out both roles."

ADAM SANDLER

Actor and comedian known for his comedy roles, has received the Mark Twain Prize for American Humor in 2023.

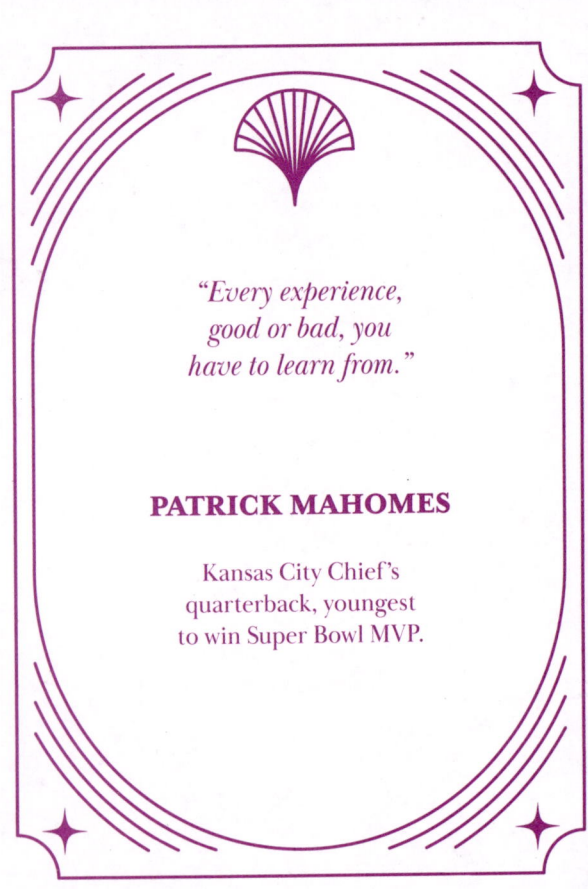

*"Every experience,
good or bad, you
have to learn from."*

PATRICK MAHOMES

Kansas City Chief's
quarterback, youngest
to win Super Bowl MVP.

"Once you get rid of the idea that you must please other people before you please yourself, and you begin to follow your own instincts—only then can you be successful . . ."

RAQUEL WELCH

Iconic actress famed for roles in *Fantastic Voyage* and *One Million Years B.C.*

"I am a collection of thoughts and memories and likes and dislikes. I am the things that have happened to me and the sum of everything I've ever done."

MACAULAY CULKIN

Actor and singer best known for starring in the *Home Alone* series.

*"The more you are positive
and say, 'I want to have a good
life,' the more you build that reality
for yourself by creating the life
that you want."*

CHRIS PINE

American actor, best known
for his roles in the *Star Trek*
reboot series, *Wonder Woman*
and *The Princess Diaries 2.*

"I don't associate with people who blame the world for their problems. You are your problem. You are also your solution."

MELISSA McCARTHY

Famous actress, screenwriter and producer, best known for her roles in *Gilmore Girls* and *Bridesmaids.*

"Love love and cherish life. Also, just eat the cake."

AARON PAUL

Actor and producer, who received multiple Emmy Awards for his role as Jesse Pinkman in the series *Breaking Bad*.

"Grief is a thing best shared."

PRINCE HARRY

A British royal family
member, army officer, is
fifth in line of succession
to the British throne.

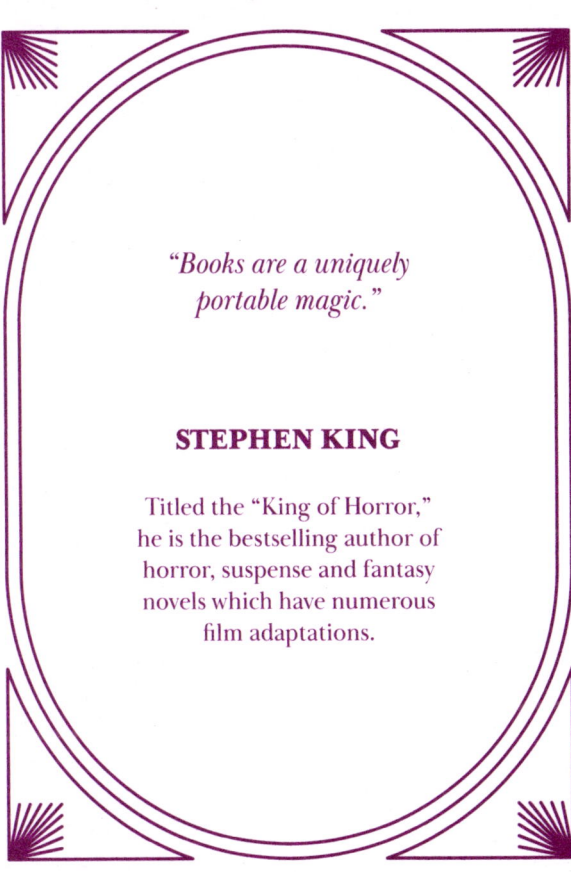

"Books are a uniquely portable magic."

STEPHEN KING

Titled the "King of Horror,"
he is the bestselling author of
horror, suspense and fantasy
novels which have numerous
film adaptations.

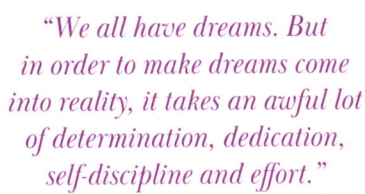

"We all have dreams. But in order to make dreams come into reality, it takes an awful lot of determination, dedication, self-discipline and effort."

JESSE OWENS

American athlete who specialized in sprints and long jumps. Won 4 Olympic gold medals.

*"Waking up in TRUTH
is so much better than
living in a lie."*

IDRIS ELBA

English actor, rapper,
singer and DJ, he is best
known for his roles in
The Wire and *Luther*.

*"If you don't mind haunting
the margins, I think there
is more freedom there."*

COLIN FIRTH

Award-winning English actor
known for playing Mr. Darcy in
the BBC *Pride and Prejudice* series
and other diverse film roles,
awarding him with accolades.

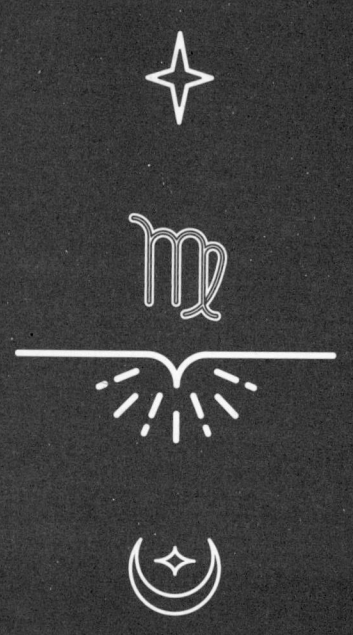

"Take pride in what you do. The kind of pride I'm talking about is not the arrogant puffed-up kind; it's just the whole idea of caring—fiercely caring."

RED AUERBACH

Basketball coach known for modernizing the sport with team play strategies.

> *"True beauty is knowing who you are and what you want and never apologizing for it."*

P!NK

Highly successful singer-songwriter known for her contralto voice and diverse music styles.

"I love seeing young people take a stance and not be afraid to say how they feel, and protest when they feel things are wrong or unjust."

ZENDAYA

American actress and
singer who rose to fame
from Disney's *Shake It Up.*

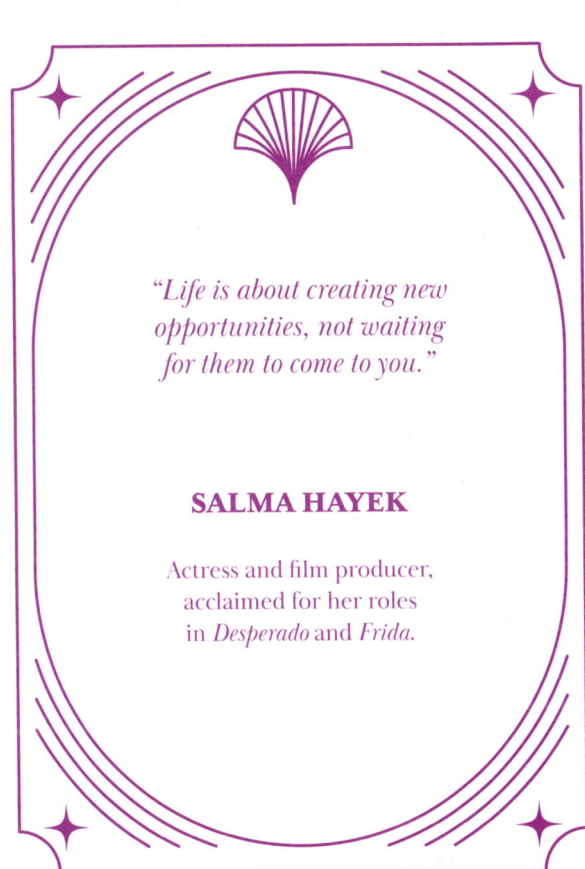

"Life is about creating new opportunities, not waiting for them to come to you."

SALMA HAYEK

Actress and film producer, acclaimed for her roles in *Desperado* and *Frida*.

"Nothing in life is more liberating than to fight for a cause larger than yourself, something that encompasses you but is not defined by your existence alone."

JOHN MCCAIN

Long-serving US senator and navy officer, who ran for president in 2008.

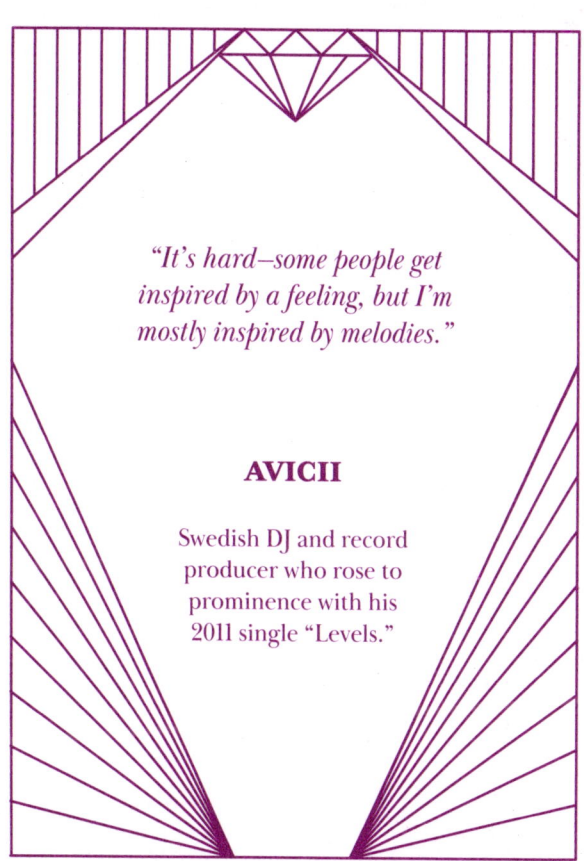

"It's hard—some people get inspired by a feeling, but I'm mostly inspired by melodies."

AVICII

Swedish DJ and record producer who rose to prominence with his 2011 single "Levels."

"I just try to be myself, whatever that is. I don't think about how I'll be remembered. I just want to be consistent over a long period of time. That's what the great players do."

DAN MARINO

NFL quarterback for Miami Dolphins, remembered for his quick release, has been recognised as one of the greatest despite never being on a Super Bowl-winning team.

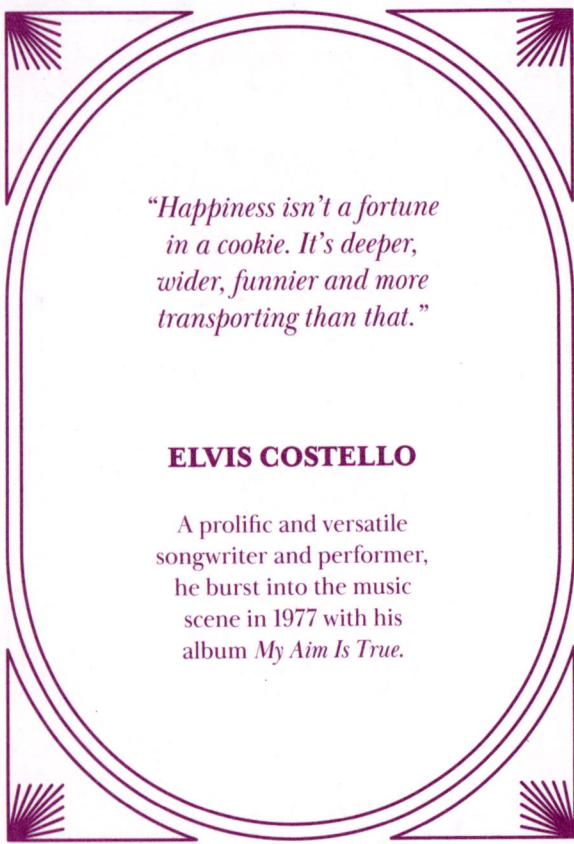

"Happiness isn't a fortune in a cookie. It's deeper, wider, funnier and more transporting than that."

ELVIS COSTELLO

A prolific and versatile songwriter and performer, he burst into the music scene in 1977 with his album *My Aim Is True*.

*"Being famous is wicked,
but it's better to be normal."*

RUPERT GRINT

English actor,
best known for playing
Ronald Weasley in the
Harry Potter film series.

*"Winning is great,
but it's not everything.
The journey and the effort
put into it matter more."*

DOMINIC THIEM

Austrian tennis player who
has been ranked world's
No.3 by the Association
of Tennis professionals.

*"I don't think that
success can be measured
by numbers, it's really
about how you feel."*

DOLORES O'RIORDAN

Irish singer and songwriter,
best known as lead vocalist and
lyricist for The Cranberries.

*"Difficult times often bring
out the best in people."*

BERNIE SANDERS

Longest-serving
independent senator,
known for bringing a leftward
shift in the Democratic Party,
ran for president in 2016.

"The whole point of team competition is to pick your teammates up."

ANDY RODDICK

Former American
tennis champion who won
the 2003 US Open singles
title among others.

"When you hear my music and you feel the emotion, it's real. When you see me in a film and you see a tear, it's real."

JENNIFER HUDSON

Singer and actress,
American Idol alumna, is
the third African-American
woman to be an EGOT recipient.

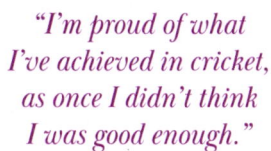

"I'm proud of what I've achieved in cricket, as once I didn't think I was good enough."

SHANE WARNE

Australian cricket legend known for his spin bowling, was a member of the 1999 Cricket World Cup-winning Australian team.

"When you are winning too much, sometimes you think you should never lose again. I am learning to lose."

GORAN IVANIŠEVIĆ

Former Croatian tennis player, renowned for being the only wildcard entry who became a Wimbledon champion.

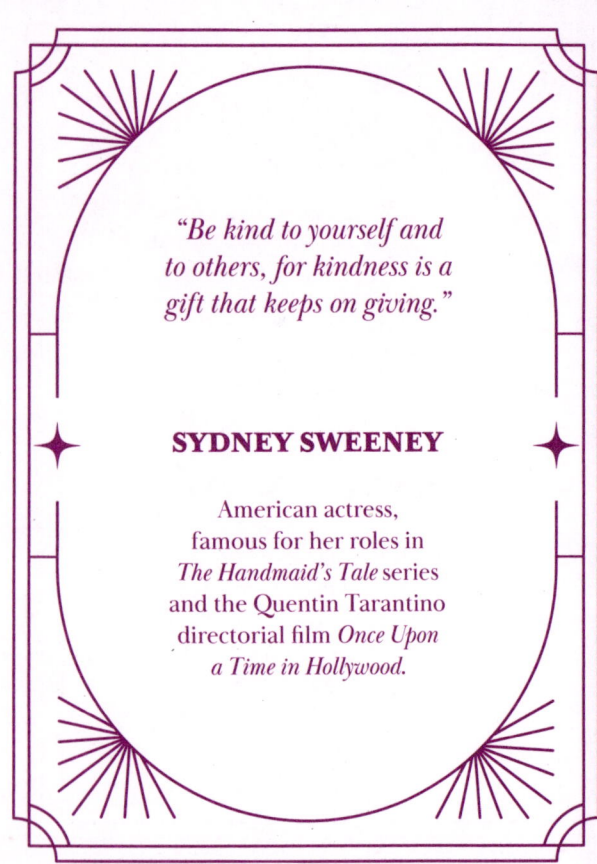

"Be kind to yourself and to others, for kindness is a gift that keeps on giving."

SYDNEY SWEENEY

American actress, famous for her roles in *The Handmaid's Tale* series and the Quentin Tarantino directorial film *Once Upon a Time in Hollywood.*